Ivana Gardiánová
Martina Helclová
Stanislav Koráb

Belgian Shepherd Malinois, German Shepherd Comparison of IPO Results

Ivana Gardiánová
Martina Helclová
Stanislav Koráb

Belgian Shepherd Malinois, German Shepherd Comparison of IPO Results

IPO - FCI World Championship and CZ Championship
2003 - 2011

LAP LAMBERT Academic Publishing

Impressum / Imprint

Bibliografische Information der Deutschen Nationalbibliothek: Die Deutsche Nationalbibliothek verzeichnet diese Publikation in der Deutschen Nationalbibliografie; detaillierte bibliografische Daten sind im Internet über http://dnb.d-nb.de abrufbar.

Alle in diesem Buch genannten Marken und Produktnamen unterliegen warenzeichen-, marken- oder patentrechtlichem Schutz bzw. sind Warenzeichen oder eingetragene Warenzeichen der jeweiligen Inhaber. Die Wiedergabe von Marken, Produktnamen, Gebrauchsnamen, Handelsnamen, Warenbezeichnungen u.s.w. in diesem Werk berechtigt auch ohne besondere Kennzeichnung nicht zu der Annahme, dass solche Namen im Sinne der Warenzeichen- und Markenschutzgesetzgebung als frei zu betrachten wären und daher von jedermann benutzt werden dürften.

Bibliographic information published by the Deutsche Nationalbibliothek: The Deutsche Nationalbibliothek lists this publication in the Deutsche Nationalbibliografie; detailed bibliographic data are available in the Internet at http://dnb.d-nb.de.

Any brand names and product names mentioned in this book are subject to trademark, brand or patent protection and are trademarks or registered trademarks of their respective holders. The use of brand names, product names, common names, trade names, product descriptions etc. even without a particular marking in this works is in no way to be construed to mean that such names may be regarded as unrestricted in respect of trademark and brand protection legislation and could thus be used by anyone.

Coverbild / Cover image: www.ingimage.com

Verlag / Publisher:
LAP LAMBERT Academic Publishing
ist ein Imprint der / is a trademark of
OmniScriptum GmbH & Co. KG
Heinrich-Böcking-Str. 6-8, 66121 Saarbrücken, Deutschland / Germany
Email: info@lap-publishing.com

Herstellung: siehe letzte Seite /
Printed at: see last page
ISBN: 978-3-659-59930-9

Copyright © 2014 OmniScriptum GmbH & Co. KG
Alle Rechte vorbehalten. / All rights reserved. Saarbrücken 2014

Table of contents

1. Introduction 4
   1.1 Aim of the study 4
2. Literature overview 5
   2.1 Belgian Shepherd dog and History of Malinois breeding 5
      2.1.2 History of Malinois breeding in Czech Republic 7
      2.1.3 Present of Malinois breeding in Czech Republic 8
         2.1.3.1 Breeding clubs 8
         2.1.3.2 Breeding according to Czech Malinois Club (CMC) a Belgian Shepherd Breeders Club (KCHBO) 9
      2.1.4 Malinois Working Ability 11
   2.2 German Shepherd dog 15
      2.2.1 History of Breeding of German Shepherd in Czech Republic 17
      2.2.2 Present of Breeding in Czech Republic 18
         2.2.2.1 Predispositions of German Shepherd Breeding 18
      2.2.3 Working Ability of German Shepherd 19
   2.3 Sport training 20
   2.4 Selection team 41
3. Material and Methods 43
4. Results 45
   4.1 Representation of individual breeds 45
   4.2. Comparison of score (breeds and disciplines) WC FCI IPO and CZ IPO 46
      4.2.1 Comparison of score (both breeds and disciplines) WC FCI IPO 46
      4.2.2 Comparison of score (both breeds and disciplines) CZ IPO 50
5. Discussion 56
6. Conclusion 61
7. Literature 63

1. Introduction

Belgian Shepherd Malinois is very vital and intelligent and has up to bursting with vitality, which plunges into all activities. His contract extension is the same as for a German Shepherd dog, from the police, through the rescue and sports training, where his drive and vitality very impresses. Compared to the German Shepherd Dog, however, is less adaptable to changes in the canine unit. German Shepherd dogs is breed most belongs to the world and is the most widely used working breed. It is one of the least demanding breed, breeding is very resilient and adaptable. It has excellent character traits, which are from the beginning of his recognition of the subject before all the exterior quality. It is the most adaptable to changes the handler and the environment, so its use is from the police and other armed forces, through the guide and assistance dog. It is used as a rescue and avalanche dog sports dog in all activities from the agility after the versatile competition and last but not least as a family dog.

1.1 Aim of the study

The aim was to summarize the history and breeding of the Belgian Shepherd dog Malinois and German Shepherd dog, to give an overview of the breeding in the present and the use of both breeds, but especially the sports training according to the IPO, to further evaluate the differences of both breeds on championship competitions - Czech Championship IPO and World Cup IPO and compare them in years 2003-2011 and for the nine year period together and in each of the disciplines. Work hypothesis – "Belgian Shepherd Malinois is achieved better scores in the competitive disciplines of the IPO than German Shepherd".

2. Literature overview
2.1 Belgian Shepherd dog and History of Malinois breeding

The Malinois is one of the varieties of Belgian shepherds, which was used before the year 1891 as working dog for farmers and shepherds, had uniform appearance in color and hair, but they anatomically praised by due to the work he had to perform (Vacková, 2008). According to the region in Europe are the sheep dogs in spite of its size. In the regions where is the existence of the wolves and the bears, who were also the main enemy for the herd of sheep, dogs were big and strong. Mostly all the sheep dogs are short and spiky haired, which is protected from the weather. At the end of the 19th century were breed in Germany, France, Italy, Belgium, Hungary and other countries of Europe by many of the sheep breeders and their dogs lived with large herds of sheep. At a time when these countries were breeds like German Shepherd, briard, beauceron, puli, kuvacz, etc., in Belgium sheep dogs holding the size of the withers of around 50-55 cm and their weight does not exceed 20 kg. The head was narrow, with a small triangular ear, located high on the head. The eye was dark, almond shaped with intelligent and "live" expression. Body was short and quadratic with light bones, with a lively and light movements and "cheerful temperament" (Bos, 1989).

In 1891, at the urging of friends and owners of shepherds in Brussels society for pure breeding these shepherds under the name of "Club du Chien de berge belge" (Belgian Shepherd Dog Club), sponsored by the "Societé Royale Saint Hubert" (Royal Society Saint Hubert) and already at this time was its "club" solid results. The first activity of the Club lay in the fact at 15. 11.1891 and was staged 117 "prototypes" of today's Belgian Shepherd Dog to the Veterinary Clinic in Curegham in Belgium. This meeting called later the referee Professor Adolphe Reul (M, 1998). Among the 117 dogs were European (mainly black, rarely brown or wildly coloring), coarse-haired (grey) and the short-haired (brown or beige, with a darker mask). Forty dogs were separated as suitable for further breeding, among them seven black with long hair. Professor Reul has recommended that only dogs and females paired with each other in a similar fur (Vacková, 2008).

The Club built a new standard, which set out that the dogs may be black, long-haired and short-haired grey with broken hair reddish-brown (Wagner, 1996). This eliminated a large number the good dogs from breeding, and their owners formed their own club (Vacková, 2008). In 1920 released under cover of Royal Canine Society Royal St. Hubert breeding rules. The cause of the losses, have been enormous valuable breeding material in the just completed World War II. In 22nd of October in 1929 were recognized black short haired dogs as a standard.

In October 1945 was again adapted after end of World War II to the standard of the 1920. On the basis of the resolution of 1966, concerning the varieties of Belgian Shepherd dog as a last and which provides for the award of the title CACIB four types was founded at the end of 1972, in Belgium, the provisions of which apply to this day: no longer are allowed to cross between the varieties.

The four varieties of Belgian Shepherd dog are as previously described as follows:

groenendalen = black, longhaired variety

tervueren = reddish brown variety, with clearing brushwood and a black mask, longhair

malinois = reddish brown, black mask, shorthair

laekenois = reddish brown, rough hair (Wagner, 1996).

The name of the short haired variety of the Belgian Shepherd dog gave the town of Malines, Mechelen. Keepers and breeders around Malines wanted from the beginning a working dog with symmetrical appearance and coat color for them was quite on second rate (Vacková, 2008). The Malinois is about its origin with Lakenois. The first shorthaired grey female named Diane was a product of grey short haired Lieske de Laeken and rough haired Voss Nasa bio de Laeken. Diane was mated to shorthaired and annealed dog named Samlo (Matušková, 1998). From this mating was a shorthaired dog named Tommy, which became the Foundation of Malinois (Pisarčíková, 2006). This dog in connection with the massive reddish brown female Cora van Optewel gave a legendary Trop one of the founders of the breed. The second line was created again male, arising from the concentration of grey shorthaired Mouche (sister named Diane) and red shorthaired Voss des Polders. The famous descendant Dewet, was reddish brown (Matušková, 1998).

In the beginning of breeding was not too much about appearance but for work use. At the beginning of the 20th century was particularly valued as a versatile working dog, and was increasingly used in army and by the police (Bossi, 1989). Soon out did the Malinois be as a working breed, all other varieties of Belgian Shepherd dog, and it has remained to present.

In the 1980´s of 20th century experienced a Malinois in Central Europe, a remarkable boom. Although in the past he knew only a few real lovers of Belgian Shepherds, it suddenly grew extremely popular dog used in sports, which aroused positive interest also for the police and customs authorities. Malinois in recent years occupy the highest positions at the World Championship FCI IPO (Vacková, 2008).

2.1.2 History of Malinois breeding in Czech Republic

The first import of Malinois came into the Czech Republic (then Czechoslovakia)at 1988 from Holland and it was female Halusetha´s Tekla (champion CZ, ZM), it was the beginning of breeding this breed. Not long after this import was imported from also from Holland dog Halusetha´s Urban (champion CZ, ZV2, IPO3, OP1). At the end of 1988 was next still imported from Holland female Sonja v. Tasca's Home (later, r. CACIB, i.e. a reserve candidate international champion, ZV3, i.e. the test of the national examination regulations of the highest degree, IPO3, i.e. the test of the international examination regulations of IPO of the highest degree, OP1 i.e.. defense test).

Since 1989, they started to write separately each of the Belgian Shepherd varieties in studbook (Czech Kennel Studbook). The first number got female Ass' Kiri, imported from Denmark (in kennel "Mateo"). The second was written female imported from Belgium named Mero du Maugré (champion CZ, ZM, IPO3, ZVV2), she worked in the Kennel "Bohemia Alke"). The first puppies were born in the Czech Republic at 11. 9.1989 in Kennel "z Hanky" by female Sonja in Tasca's Home and the male Halusetha's Urban (Vacková, 2008). In the litter were five males and three females and in further breeding in the Czech Republic have been used three of them (two females and one male: Agatha - IPO3, OP1, founder of Kennel "Malidaj"; Asuntha - ZV2, IPO2; Attol - ZV3, OP2, IPO3, SP1, ZZP2, ZLP2, participant of the championship CZ, Wold Cup, etc.) (Pisarčíková, 1997).

The second litter was born 8. 10. 1989 at "Hanako" Kennel, female Halusetha's Tekla and the Austrian's male Ramus du Colombophille. The most famous descendant from this litter was female Audi Hanako (ZV1, IPO3) founder of Kennel "z Huckelovy vily" (Vacková, 2008). Since then many Malinois were imported from Holland, Denmark, Belgium, Austria, Italy, France and other countries and breeding of Malinois in Czech Republic was on a very high level (Pisarčíková, 1997).

2.1.3 Present of Malinois breeding in Czech Republic

2.1.3.1 Breeding clubs

Currently there are two clubs for Belgian Shepherds in Czech Republic. The first is a Belgian Shepherd Breeders Club (KCHBO) and the second Czech

Malinois Club (CMC). In 1989 was issued the first club newsletter of Club Breeders of Belgian Shepherds (KCHBO) which was established in November 1988 in Prague. This club became in the year the only 24. dog's special club registered under ČSCH (Czech Breeders Association). The club counted with poolingo of all varieties, although at the time has not been any Malinois litter and no dog of the variety Lakenois in Czech Republic. Since 2002, brings together the club KCHBO also breeders and fans Australian Shepherd dog.

In early 1992, has established a competitive Club Berger Belge. Over time, the only place in the ranks of the Malinois and the member owners meeting in 2006, the Club Berger Belge was renamed to Czech Malinois Club (CMC) and continued to be dedicated to breed a variety of Malinois only (Pisarčíková, 2006).

2.1.3.2 Breeding according to Czech Malinois Club (CMC) a Belgian Shepherd Breeders Club (KCHBO)

Between the basic conditions for the inclusion of a male or female in the breed include:

a/ purebred a certified pedigree issued by the studbook recognized by the FCI and the Club, in the case of imported individuals condition re-register

b/ hip dysplasia done with marked degree of Dysplasia in the certificate of origin, with the highest permitted level for inclusion in breeding is 2/2

c/ descriptive compliance shows and temperament test (or performance tests) with the result - succeeded

d/ reach minimum age the boundaries for the breeding, which for dogs and bitches is 18 months

e/ DNA profile for all individuals used in breeding from the 1. 7. 2011.

Inclusion in breeding for the Belgian Shepherd Malinois:
a/stud-satisfied

- the result of HIPS RTG oft a maximum 2/2
- participation in a "Descriptive show" - with the result of no "decommissing" defects according to standard from 15 months of age the dog
- examination of the nature - either in the form of temperament test with the result of "Stud" (it is possible to repeat) from 15 months of age of the dog, or the composition of the test, SchH1, ZVV1, IPO1 with the result of "succeeded".

Males or females were after resulting placed among the breeding individuals as "Stud".

b/ increase breeding class - working males - exams: IPO3, SchH3, ZVV3, females - exams: IPO2, ZVV2, SchH2

- demonstration of the defense, i.e.: the spillway from the mock inspection performance. The defense is implemented, as passed bonitation of German Shepherd. Increasing the employment of breeding is carried out once a year.

- exterior - breeding base, the title of Champion of the Czech Republic and 1x CACIB (candidate international champion)

c/ non breed - males or females, who do not fulfill the conditions for inclusion in breeding.

The breeding according to the conditions for the inclusion of KCHBO males or females for breeding:

a/ pedigree issued by the studbook recognized by FCI and Czech Moravian Cynological Union, in the case of imported individual re-register in the Czech studbook

b/ minimum - 18. month of age

c/ evaluation and examination for Dysplasia of the hip joints, with the result of 2/2, inclusive. In the case of evaluation of your organization is permitted a maximum result of OFA organization.

d/ met a descriptive show with the result of "Stud" and a temperament test.

e/ with effect from the 1. 7. 2011 must have males and females use in breeding DNA profile according to the standard ISAG2006.

Character test of KCHBO:

Temperament (character) test shall be carried out in order to assess the nature of the Belgian Shepherd dog, with the greatest emphasis is placed on a balanced, non-conflicted behavior of well socialized dog. The ideal dog is here understood such a dog whose temperament is properly balanced, confident. The dog has a neutral or friendly relationship to people and animals and has a good link to his handler (Pisarčíková, 2006).

2.1.4 Malinois Working Ability

Malinois is perfectly equipped for any kind of training. His excited docile, the ability to quick understanding, it does not occur to him that so many hip dysplasia, speed, and agility - all of the above properties make it an ideal partner (Wagner, 1996). The original mission as shepherd dog to keep and keeping the herd is not much nowadays. At the present time is for the breed's important application in the professional training of the armed forces and sports training (national order), IPO (international test procedure) and SchH (German examination regulations). In the customs service achieves excellent results for its excellent and unerring sense of smell, agility and robustness. His lighter body weight allows him to penetrate almost everywhere. When searching for explosives and drugs recorded great achievements. The police in Europe use the Belgian Shepherds Malinois even for emergency units, the guide the blind people and physically handicapped. Despite his temperament is suitable for this training. An important application is like a dog guard and family (Matušková, 1998).

When evaluating the disciplines (defense, track, obedience and more), it was found that males are better than females. The evaluation also found that Belgian Shepherd Malinois is the best for work from all varieties of Belgian Shepherd dog (Courreau and Langlois, 2005).

Rescue training – Union of Cynological Rescue Brigades (UCRB) of the Czech Republic is a social organization engaged in the rescue work, using specially trained dogs. Their training is conducted to search for the living and dead persons in a many types of environments. In the winter season it is all about finding tucked people in the snow or in a landslide. In the summer they search for people fight and lost in the inaccessible terrain, mostly children and the elderly. A specific chapter consists of the above - mentioned people, search under water with the help of boats, on the bow lies a dog by sniffing - all the surfaces and locates drowned. Union is divided into regional brigades, which operate on the territory of the individual regions. Their activity is managed by the Presidium of the Union elected once in 5 years that is between the meetings of the executive body. Of the currently best trained members of the Union is prepared and continually Standby unit UCRB that is ready at any time, at the invitation of subsequent intervention not only in the Czech Republic, but also abroad. In the emergency unit of the Czech Republic is Union also a Malinois named Fany Malidaj Mrs. Hany Klímové (Sedlák, 2007).

Dog dancing is the presentation of the perfect relationship between handler and his dog. It is a remarkable demonstration of the exercises of fantasy and music. It is the show, which entertains audience of all ages. Club Dance with a dog in the Czech Republic is a member of the FCI. It is a mixture of exercises, the classic of obedience, revolutions stepper variations, and jumps. The dog moves forward, to the side is reversing before the handler, winds its way between her legs, alternately lifting paws, worships, etc. it's all in the rhythm of the music. Handler dressed in appropriate costume underlining genre of music, accompanied by dog dance movements and a sort of regulating the movement commands the dog into well-rehearsed choreography. The overall appearance is on the viewer so that she has the impression that the dog really dances (Malíková, 2008). Malinois appear in competitions such as: Aralia Novterpod, Abate Novterpod, Achala Novterpod.

In **mondioring** is definitely not training dril and stereotype. The greatest difficulty and demands of the mondioring in contrast, lies in its infinite variability, and creativity. It's not just about art together and the quality assessment of the training level of their charges, but primarily on the examination of the dog disposition. The handler does not exercise the option desktop training and enter the dog to her at the time of the initiation of the test. Ring (exercise area) is also thematically focused. To implement defenses are available two helpers, clad in full body suits for biting. Have the responsibility to examine the nature and perfectly psychological resistance of the dog. In the mondioring there are three degrees of difficulty according to the tests. The test can be met only by means of the race. The trial of the mondioring was approved in September 1994. The elements basis was based on the long standing traditions of the French and Belgian ring. In these countries this training has been dedicated since the beginning of the last century (Vala, 2007). In Czech Republic involved mondioring mainly in the Kennel of "Huckelovy Vily" married couple Volný and Mrs. Gondeková.

Agility is a demanding competition skills, the ideal exercise growing prowess of handler and animal. In addition, it requires and psychic abilities. Dog gets confidence and speed of reaction. Mainly however it is important to create a perfect synergy between the pair, which is indeed a fascinating when agility races. In 1977 she appeared for the first time as an attractive agility fill breaks in the Cruftofs exhibition in London the world's largest dog show. It was the use of dogs inspired by horse jumping. This is sport with remarkable speed, spread across the world and won fans everywhere. In 1991 the FCI has developed the binding rules. It was supplemented, revised and adapted to best suit practice. The new rules of FCI came from 1. 1.1996. Base jumping is composed of 12 to 20 different obstacles (Wegmannová, 2003). The basic aim is to overcome the agility track flawlessly and jumping in the best time. Show jumping is a bounded space, which are under construction obstacles and the order of the judge. Show jumping may take the form of agility (where they are under construction and the so-called zone barriers-balance beam, "A", swing), or dumping (where the zone barriers may not be). The obstacles are numbered and it is necessary to overcome them in the right way in the right direction and order. For the mistake, as well as for the refusal or overcome the maximum time penalty (Divišová, Podešťová, Benda, 2003). In the Czech Republic agility is widely developed sport.

Flyball is a sport and entertainment for lively dogs of all breeds. Compared to the agility has the advantage that the handler is not so demanding, so he can pay and physically less equipment handlers. It features two of four teams in the two same flyball rail systems. Each track consists of four of the same jump obstacles and the flyball box that shoots the ball. The command handler which however remains behind the starting line, the first dog on the runway to surpass the four jumping obstacles jumps up on the plate of boxing (performs the so-called swimming round) - this starts the drive, caught one ball and retrieving it back over the obstacles to adoption. At the moment when the first dog of the team crosses the finishing line juts out another dog. The team that wins is the first goal of all four dogs (and balls). The interplay of cooperatives and mutual tolerance in flyball dogs is very important (Vyskočilová, 2008).

2.2 German Shepherd dog

Breeding of German Shepherd dogs in Germany has a long tradition. Already in the oldest extant in "Gestner" law of 6th century states, that "anyone who kills a dog that is able to protect the herd, the hard punished." At that time it was the dogs difficult, capable of confronting the big predators. With breeding dogs similar to today's sheep dogs, therefore medium sized and strong, started at the end of the thirty years ' war, when it defined the boundaries and consolidate so general safety - i.e. approximately at 17th of the century. The packs of wolves and bears were almost eradicated in Europe. Therefore, the danger from them is almost there and the main task of the shepherd dogs which were kept the herd in defined areas. This job required a lighter type dogs, moving and relentless. A consistent appearance and is similar to the construction of the body and the dogs were very different in different regions.

Only from the middle of 19th century began to prevail as sheep dogs middle and south German - Duryn and Wűrtenberg. The first of them usually the wolf grey colors right ears, but often his tail coiled or spiraled. The second have black characters of different shades mostly tipped ears, but properly borne by the tail (Stibůrek, 2002). Efforts for unification of the two types of herding dog, and the creation of a new comprehensive breed continued to the founding of the first Association for the Breeding of the German Shepherd dog named "Phylax" (16 December 1891), which is however considerable differences in disunity in 1894, crumbled. The idea, of course, already taken and the increasing need to shepherd dogs gave rise to a new organization of the Verein fűr Deutsche Schäferhunde Der (SV), namely the Association for German Shepherd dogs from which later became the world's largest organization that deals with just one dog breed. It organized a young cavalery officer and important cynologist Max von Stephanitz, and his prematurely dead friend Artur Meyer. He should have a clear idea about how the new national breed of shepherd dog look and what must have properties, but what was most important - have managed to convince breeders and fans of these dogs (Soukupová, 2006). The day "D" is known as the day of 22. April 1899, Max von Stephanitz demonstrated their dog Hector von Linkershein at the exhibition of dogs in Karlsruhe. The official name of the dog, which was later entered into the studbook, as a forefather of the breed, Horand von Grafrath sounded. Even on the day of the exhibition was SV based (Metzová, 2007).

In his famous book a "German Shepherd in Words and Pictures", writes the following: "Stephanitz, for enthusiasts of that era was the epitome of Horand and completing their most daring dreams, was the great - his height was around 61 - 62 cm (which would even from today's perspective was a solid medium height), had strong bones, beautiful lines and noble shaped head, purely built and muscular body was full of energy. His character was in accordance with the quality of its external appearance, excellent in obedient devotion to his master, he had a straightforward nature. Horand forward these fabulous properties in its other descendants "(Allan and Allanová, 1997). Stephanitz in search of the modern form of the German Shepherd dog came from rather diverse breeding material quantity of types of shepherd dogs from the extensive territory of the German Empire. The exterior was mainly the result of balancing the emphasis on performance (Císařovský, 2011).

Over the years through the breeding of German Shepherd dogs to different periods. The appearance is typical of dogs and a stable and improving. Expanded number of blood lines. The most prominent males, which had the greatest impact were in addition to the aforementioned Horand also males named Klodo Boxberg, Rolf v. Osnabruckerland, Vello von Skrben Faulen, Canto and Luando von Wienerau, from the last period Uran von Wildsteigerland. It goes without saying that in 100 years in hundreds of excellent breeding males and females, who in the development of the breed hit. The dogs however can be described as a "generational", which bordered certain developmental periods (Stibůrek, 2002). Winning the onset of German Shepherd around the world was up to then when his excellent truly unique features such as the dog of the staff. Soon he became popular, fashionable dog, who, with his intelligence, grace lines and other valuable properties, especially fidelity and watchfulness, became "darling" for breeders around the world (Najmanová and Humpál, 1981).

2.2.1 History of Breeding of German Shepherd in Czech Republic

Breeding of German Shepherd dogs have a long standing tradition in the Czech Republic. Already in 1912, was founded in Brno branch of the Austrian Association for the Group of Moravia. After the decline of Czechoslovakia was formed, even though the German the first Association for German Shepherd dogs in Czechoslovakia based in Brno. First time were apparently exposing German shepherds in 1905 on the first provincial police and commercial exhibition of dogs and were here three individuals. Since 1923 have been German Shepherds entered into the first Czech Studbook and led by Czechoslovak Union of the Cynology. After the first World War was breeding a German Shepherd a little advanced, but already in the 20th century (years of the last century), however many individuals which were on the level – used to the exterior and work.

The highest boom of reached breeding in Czech Republic was during the second World War. It was due, in particular, easier import opportunities or rentals of dogs from Germany from abroad. After World War II left a lot of good dogs, but with regard to the economic situation leaved activities many breeders. In 1950, it was imported from Germany for several breeders of German Democratic Republic and by breeding for us getting into the impact of this breed. Before November 1989 had just a few breeders the ability to import dogs from German Federal Republic. The quality of dog importsed were different, but their blood lead was the basis of later offspring, which could continue. It was then just a matter of opening of borders and the possibility of protection abroad to make our breeding on the superior level, which we can observe in the past few years not only on our exhibitions but in particular by placing our puppies as well as on foreign exhibitions, particularly in the country of origin, in Germany (Šiška and Jánský 2006).

2.2.2 Present of Breeding in Czech Republic

2.2.2.1 Predispositions of German Shepherd Breeding

The German Shepherd is in the Czech Republic organized by the Club of Czech German Shepherds (ČKNO). This Club is incorporated through the Czech Kennel Union (CCP) in the Czech-Moravian Union Canine. ČKNO is a member of the World Union of German Shepherd Clubs (WUSV) which are based on the training requirements of breeding and training of the breed (Šiška and Jánský 2006).

The conditions for the inclusion of a German Shepherd for breeding:

1. exhibition with at least a "good" award from the adolescents or adults from special exhibitions of German Shepherd

2. performance test of at least 1. degree (SchH1, ZVV1, IPO1)

3. RTG DKK (x-ray exam of hip dysplasia) a maximum of 2/2

4. age of 18. months

5. valuation – to get called selection class.

6. DNA - from 1. January 2009 must have an individual removed for breeding a blood sample to determine the identity of the (DNA) - record in studbook.

For German Shepherds with the valuation carried out two times. In the selection of the individual is included in the period of two years, for its renewal must be demonstrated once again passed on bonitation (Soukupová, 2006).

Valuation is the valuation of the noncompetitive individuals. Judge each dog (bitch) assesses how the exterior, so from the perspective of nature.
Character part:
1. passage of a group of persons
2. the spillway of the impactor
3. the control of performance.

In the end of the validity of the first bonitation, may participate in the evaluation again and dogs if they fulfill all the conditions laid down for the test nature, are included in the selection of breeding for life. Individuals which are not in the selection of breeding, i.e. individuals in controlled breeding, breeding are under 8 years of age. The same applies to individuals who did not rebuild the valuation or have not met the required conditions valuation (Novák, 2008).

2.2.3 Working Ability of German Shepherd

One of the first role was the role of a German Shepherd as police dog. In 1903 were recommended into the police forces use of the shepherds. Began to test their abilities and the results were satisfactory enough that the police organizations were recommended to try and take the dogs as an integral part of the system of law enforcement. The Government has established in Grunheide near Berlin, breeding and training center for police dogs. Until the outbreak of the first World War travelled to Germany police groups from many countries around the world and returned home with great trained dogs.

So were the foundations of sections of the police dogs in the international scale (Allan and Allanová, 1997). In the first World War was German Shepherd used to search for wounded soldiers, throwing his equipment for first aid phone cable, he served as a guard of the stockpile. After the war, Germany began to train German Shepherds newly for blind soldiers, around 1950 was a shepherd's guide dog in the world (Fogle, 1996). The use of working breeds of dogs for the armed forces to patrol activities, detection of explosives and drugs are in the current world political climate, thanks to the active areas of research. Repeated findings of these researches confirmed breeding and gender differences in the use for these purposes (Sinn, Gosund, Hilliard, 2010).

It is currently the most used breed worldwide, the German Shepherd breed for working purposes, armed forces through rescue, Braille and assistance dogs to all dog sports, where achieved great score (Trávníčková, 2005).

2.3 Sport training

In the time before 1989 were under the auspices of Svazarm (Association for Cooperation with the Army) based Cynological training ground and explore (basic Cynological organization), which in large numbers had endured to this day. In the case of the necessary protection of the motherland had to be here to exercise dogs actually serviceable. It also demands the universal training. Over time the versatile training were shifting increasingly to sports focus, where the big competition taking place on stadiums, which is not using in practice much in common. Emphasis is placed on a joyous and accurate implementation of exercises with complete maneuverability (Soukupová, 2006).

The exam regulations of the IPO

The test of this international trial of the IPO (international trial rules) has been processed by the FCI Commission for Cynology. Adopted and approved by the FCI was 3. April 2011 in Rome and is valid from 1. January 2012.

The exam order was compiled and processed by the Commission in the German language. In the case of confusion that could arise in the translation into other languages, is a crucial text in German language.

This test is valid for all the Member countries of the FCI. Must follow all the action with the content of international examination degrees (and competition) (Šveráková and Hodek, 2012).

IPO-3 is divided into:

section A 100 points,

section B 100 points

section C 100 points

for a total of 300 points.

IPO 3 SECTION "A"

The foreign section of the track, at least 600 paces long, 5 sections, 4 quarries (about 90 degrees), 3 articles, at least 60 minutes old, time to prepare 20 min. tracking 80 points. Subjects (7+7+6) with 20 points, total of 100 points.

General provisions: the judge or the person responsible for the stretch of track determines the shape of the tracks due to the area, which is intended to stop laying. Objects or quarries may not be on all tracks laid in equal distances from each other. The beginning (footsteps) tracks must be well marked with the label, which is in the immediate vicinity left inserted into the country. The order of the participants is the referee after laying the tracks draw. Setter of tracks shown objects before laying tracks to arbitration or the person responsible for the stretch of tracks. Only objects can be used very well heavy with the smell of (at least 30 minutes).

Setter of tracks from the stance briefly, and then it is a normal step in the specified direction. There are also quarries are a normal step, first the subject is seen after at least 100 steps 1. The second section, subject to the 2. or 3. section, the third course at the end of the tracks. Items must be placed on track for walking. After put the last subject is going the setter on a few more steps in a straight line. In the course of one track must be considered different objects (e.g.: leather, textiles, wood). Items may be the most 10 cm long, 2-3 cm wide and 0.5-1 cm thick, and its color differ significantly from the surrounding terrain. All items must bear the number is identical with the number plate indicating the beginning of the track. After a period of put the tracks must be the handler and dog out of sight. The referee, setter and the accompanying person must keep the dog during the work in the space, in which the pair (the handler and the dog) the right track to prepare (e.g. not the handler to prevent).

    a) audio signal: the "search" or "footprint". An audio signal for search is allowed at the beginning and after the first and second subject.

    b) implementation: prepares the dog's handler to work on the trail. The dog can search either freely, or 10 m long cord. Tracer cord 10 m long can be conducted through the rear of the dog, side or between the front or rear of the limbs of a dog. It can be mounted directly on the collar of a dog that is not set to download or place designated for tracker harness. Thoracic or lumbar are allowed without additional harness belts. The challenge of reporting the dog handler in the basic position of the arbitration and shall notify him whether dog items rises, or indicates. Before the track, in the course of marketing and when the total work on the trail must refrain from any kind of pressure on a dog. On the instruction of the arbitration is the dog slowly and quietly drafted at the beginning of tracks, and is on the track list. The stance has dog smell traces of hard, low nose and sniff. Then you must also monitor the progress with interest tracks low nose and uniform pace. Handler followed by a dog at the end of the cord with the passing of the tracer 10 m free tracking. Also preserves distance 10 m from the dog. If the handler holds the cord in his hand, the tracer may be this SAG. Quarries must prepare the dog for sure. After their establishment must dog continue at the same pace in track. Once the dog finds the subject, it must be lifted immediately without influencing the handler or conclusively indicate. When lift it, can stay with him, to sit with him or come to the adoption (or alternately). Is incorrect, if the dog is raised the subject further on the trail with him continues or lies down. Refer to the lying down, sitting or standing (also alternately). If the dog handler, put described the item tracking line and goes to him. By raising the subject, it appears that the dog was found. Then picks up the tracer and continues with the dog at the same place in track after their brush with found objects show arbitration.

c) rating: If you track produced with interest, uniformly and persuasively, and the dog seems to be a positive relationship to its search, speed is not the procedure evaluation criterion. Authentication, without the dog track left, it is not an error. Wrong track, high nose, defecating in the verification of a dog, the rings on the quarries, long, hesitant, lifting objects, providing assistance, the not laying of the subject is punished by the loss of points. If the handler moves away from the tracks more than the length of the cord, is the tracer work the dog on the trail ended. If they leave the dog track and the handler is being held, is invited to referee the handler of the dog followed. If this challenge is not obey, it is the arbitration work dog on the trail ended. If it is not within 20 minutes after putting the dog to heel strike reached the end of track, is the work of a dog on the trail of the Arbitration also ended. Points earned until the end of the tracks are specified. For no lifted or unmarked objects, points do not. Alternate lifting and marking objects on one track is incorrect. Method for reporting the handler applies lifting or markings on the trail of the no put course is rated within the section. The distribution points for tracking on sections must correspond to their length and degree of difficulty. Evaluation of individual sections is carried out marks and points. If the dog does not appear on the track of interest (the longer the dog at the same place without the efforts to search for tracks), the track may be terminated, even though the dog on the trail is located.

IPO 3, SECTION "B" Exercise

General provisions:
Judge gives instruction to start each practice. All other performances, such as turnovers, stop, change the way you walk, etc. the arbitration shall be carried

out without any instruction. Sound commands are listed in the guidelines. They are normally spelled short commands. Can be pronounced as in any language for one and the same activity must be requested but always the same. If a dog fails to exercise or even on the third part of the audio signal this exercise is not rated. Instead the command for calling the dog's name may be used. The use of the dog's name along with an audible command call is rated as the second in command.

In the basic position of the dog just sits right next to the left foot and the handler so that the scapular was at knee level handler. Each exercise begins and ends with the basic attitude. The final position of the base of each practice is at the same time the input base position for the following training, relocation is not required if the dog handler. Preoccupation of the basic attitudes at the beginning of practice is permitted only once. Short applause is allowed only after their practice from the basic position. After its implementation is a fundamental position of the new handler. In any case between the praise and the beginning of the new practice must be maintained a distinct pause (approx. 3 seconds).

After the basic position which is followed by the implementation of the commitment to exercise. After at least 10 and a maximum of 15 steps gives the handler a dog audio signal for its implementation. Between the "switching of seat in" and the command to sitting, as the arrival of the handler to delayed the dog must be maintained clear breaks (approx. 3 seconds). The handler can be accessed from the front or to the rear of the dog. Manageability of a dog without a leash must be demonstrated even at crossings to other exercises. When leaving the handler for the dumb bell as the dog with him. To accept it dog handler or playing with a dog is not allowed.

Turnovers are realized by the handler on the left. These exercises may be a dog to the leg of a dog handler to receive the circumvention or from the front, however, always in the same way throughout the test. After the "for sitting" can take the basic position of the dog at the foot of its circumvention or handler from the front, however in the same way throughout the test. Hurdle has a height of 100 cm and 150 cm width. Leaning wall consists of two at the top of the United wall 150 cm wide and 191 cm tall. At the bottom are the two walls apart from each other, so that the total height of the obstacles was 180 cm, oblique walls the whole area must be covered with non-slippery material. The walls are in the upper part with 3 slats 24/48 mm. all dogs on one event must use the same obstacles.

In the implementation of aport/retrieve is allowed, to use only wooden retrieve dumbbells-hereinafter referred to as objects (retrieve free 2000 g weight, retrieve and big weighing 650 g). Dumbbells prepares organizer and must be used by all participants. In the implementation of retrieve is not allowed to advance to the dog into the muzzle. If the handler fails to perform one exercise, is called arbitration, without the loss of points for its implementation.

1. without handler manageability, 10 points

a) sound command: "leg" Sound command is enabled only for moving handler and the change of the way of walking.

b) implementation: the handler the dog goes to the referee, the dog sits down and give him the message. From the basic position of the dog to an audio signal must "stay" handler, joyfully and carefully in a straight line to follow. It must always be located on the left side of the handler, the scapular always at knee level handler. When you stop the dog must independently, quickly and in a straight line, sit down at the beginning of practice as a dog handler, without stopping, 50 steps straight, after a turnover by about-face and other 10 to 15 steps will trot and slow walking (always at least 10 steps). The transition from running into the slow walk must be carried out without any intervening steps. The individual changes to the way walk must clearly differ from each other at the speed of its implementation. In a normal walking is done at least one turnover right or left and rear. In a normal walking is made at least one stop.

During the first walk handler direct direction is in the interval 5 seconds twice fired (6-9 mm caliber), from a distance of at least 15 steps from the dog. The dog must maintain indifferently to the gunfire. At the end of practice, on the instruction of the arbitration, the dog handler passes a group of at least four of the moving persons. The dog handler is in the group, one person and one person are from the right and from the left, and bypasses at least once, stops here. Is on the arbitration, whether you allow the exercise again. After leaving the group with a dog handler occupies a primary position.

c) rating: Shoving, biasing dog to the sides, delaying the dog, additional audio commands, help the body, inattention, interfering with, or harassment of a dog handler the dog is rated points.

2. postponement of the sitting position for the march 10 points

a) after one song goes: "stay", "sit".

b) implementation: the handler the dog moving from a basic position in a straight line. After 10 to 15 steps, dog on the sound command "sit" quickly and in a straight line fits, without interrupting or changing the way the handler walking or looking back. After about 30 steps with the handler stops and immediately turns to reach the dog. On the instruction of the handler goes back and stand up to the dog.

c) rating: errors in execution, slow subsidence, restless and inattentive to postpone sitting are assessed points. If the dog's place he lies down or remains the State removed him 5 points.

3. postponement of put down for the run with 10 points and calling for assistance

a) after one song goes: "stay", "Down", "to me", "stay".

b) implementation: the handler the dog moving from a basic position in a straight line. After 10 to 15 steps a normal walk, followed by further 10-15 steps in the canter. After the dog on the audio signal to the postponement of the fast and lie in a straight line, without interrupting or changing the way the handler run or looking back. After the other, about 30 steps in a straight line to the handler stop and immediately turn to the dog. On the instructions of the command "to me" or the name of the dog is the dog sit to him. The dog has come joyfully, fast and straight, just and right to sit in front of the handler. The audio signal "to leg" dog's gathering quickly and directly next to the handler.

c) rating: errors in execution, slow getting restless postponement, slow arriving, or procrastination with the advent of, changing the position of the handler, the error in the for sitting and at the end of practice are assessed points. If the dog's place getting sits down or remains is removed him 5 points.

4. postponement of the standing up for the run with 10 points and calling for assistance

a) after one song goes: "stay" (2 x), "Stop", "to me".

b) implementation: the dog handler out of position in a straight line. After 10 to 15 steps, dog on the audio signal "stop" immediately and in a straight line, the State remains without interrupting or changing the way the handler run or looking back. After about 30 steps with the handler stops and immediately turns to the dog. On the instructions of the referee, the audible command "to me" or the name of the dog handler summons the dog. The dog has come joyfully, fast and straight, just and right to sit in front of the handler. The audio signal "to leg" dog's gathering quickly and directly next to the handler.

c) rating: errors in implementation, failing to stop after the command, the troubled standing, moving, slow arriving, or procrastination with the advent of, changing the position of the handler, the error in the for sitting and at the end of practice, are evaluated by the loss-making points. If a dog instead of standing, sits or lies down, he removed 5 points.

5. retrieve on free 10 points

a) after one song goes: "Retrieve", "Drop", "stay".

b) implementation: from the basic position of the handler discards the subject (weight 2000 g) to a distance of about 10 steps. Handler gives up the audio signal to bring when the subject is calm. The dog, who calmly sits next to his handler, the sound command "retrieve" the fast and direct way to run to the subject to grasp immediately, and handler it in a fast and direct way to bring. The dog sits tightly and directly in front of the handler and keeps the subject in its mouth, quietly, until his handler after about 3 seconds the audible command "drop" removes. After remove the sticks upright is in the right pocket, the handler's arm. The audio signal "to leg" dog's gathering quickly and directly, to the left next to the handler. In the course of the entire practice may not leave the handler.

c) rating: errors in the implementation of the basic position, slow running dog for the subject of his errors, gripping, slow bring customer, dropping the subject, playing with him, or his biting, change the position of the handler, the error in the sitting and at the end of practice are assessed-points. Also loss-make points of subject are judged too short a distance, as well as assistance dog handler, although this leaves the station. If the handler leaves the post before the end of the exercise, the exercise is rated poorly. If the dog is not the subject, the exercise is evaluated 0 points.

6. retrieve leap (100 cm) and 15 points

a) after one audio command: "forward or hop", "retrieve", "drop", "stay".

b) implementation: the handler shall deliver the dog basic position at least 5 steps in front of the obstacle. From the basic position of the handler discards the subject (with a weight of 650 g) over 100 cm high barrier. Sound command to jump, are gives when the subject is quietly situated. The dog, who calmly and is not on the leash sitting next to the handler, the sound command "hop" and "retrieve" (a sound command "retrieve" must be given during the jump) jump to overcome an obstacle, the fast and direct way to trick the subject immediately grab, and jumping to overcome an obstacle, and handler it in a fast and direct way to bring. The dog sits tightly and directly in front of the handler and keeps the subject in its mouth, quietly, until his handler after about 3 seconds the audible command "drop" removes the item after removing the sticks upright in the right pocket, the handler's arm. The audio signal "to" dog's gathering quickly and directly, to the left next to the handler. In the course of the entire practice may not leave the handler.

c) rating: errors in the implementation of the basic position, jump and run dog slow for the subject, his grasping errors, slow backward jump, dropping the subject, playing with him or his biting, change the position of the handler, the error before the settlement and their practice are evaluated corresponding to loss-making points. If a dog touches when you jump the obstacles may be removed and 1 point for each jump, when the rebound of an obstacle up to 2 points.

Partial evaluation of exercises is possible only if his three parts (jump there - bring customer subject - jump back) are at least two parts made.

* Without the glitches made jumps and bring customer subject = 15 points to Jump back there or not executed, the subject without the glitches brought = 10 points

* Jumps back and forth carried out without glitches, the subject of not given = 10 points if the course considerably after a party or is it visible, the handler without loss of points, after the arbitration request, or at its direct instruction, give up the subject again. The dog remains sitting. Help a dog handler, even though it leaves the Habitat is evaluated-points. If the handler leaves the post before the end of the exercise, the exercise is rated poorly.

7. retrieve with climbing (the leaning wall 180 cm) and 15 points

a) after one audio command: "forward or hop", "retrieve", "let go" and "stay".

b) implementation: the handler shall deliver the dog basic position at least 5 steps in front of the sloping wall. From the basic position of the handler discards the subject (with a weight of 650 g) across the sloping wall. The dog which is calmly sitting next to the handler, the sound command "hop" and "retrieve" (a sound command "retrieve" must be given in the course of overcoming the slanted walls) to overcome the big sloping wall, fast and direct way to trick the subject immediately grab a big subject, and overcome the sloping wall back, and handler it in a fast and direct way to bring. The dog sits tightly and directly in front of the handler and keeps the subject in its mouth, quietly, until his handler, after a break of about 3 seconds, an audible command "let go", the subject is taken after dog remove the sticks upright in the right pocket, the handler's arm. The audio signal "to" dog's gathering quickly and directly, to the left next to the handler. In the course of the entire practice may not leave the handler.

c) rating: errors in the implementation of the basic position, slow climbing and running dog for the subject of his errors, grasping, climbing, slowly dropping the subject, playing him or his biting, change the position of the handler, the error in the sitting and at the end of practice are assessed points.

Partial evaluation of exercises is possible only if his three parts (rope there-bring customer subject-climbing back) are at least two parts made:

* without the glitches made climbing and bring customer subject = 15 points

* climbing back there or not executed, the subject without the glitches brought = 10 points

* back and forth climbing implemented without glitches, the subject will not bring = 10 points if the course considerably after a party or is it visible, the

handler without losing any points after the arbitration request, or at its direct instruction, give up the subject again. The dog remains sitting. Help a dog handler, even though it leaves the habitat is evaluated points. If the handler leaves the post before the end of the exercise, the exercise is rated poorly.

8. broadcast forward with putting 10 points

a) after one song goes: "forward", "Down", "sit".

b) implementation: from basic attitudes as a dog handler in the commanded direction. After 10 to 15 steps gives to dog a sound command "forward" the current single order arms and myself will remain standing. After this the audio get dog command direction quickly and directly away at least 30 steps from the handler. On the instruction of the handler gives the dog the command "down" sound, which must immediately take the dog. The handler may leave raised his arm in the direction mandated by sending the dog until it is getting. On the instruction of the handler goes to the dog and stands to his right side. After about 3 seconds, at the direction of the referee, gives the handler a dog sound command "sit", to which the dog must sit down quickly and directly.

c) rating: dogs in the implementation, follow the dog handler after his maneuver, slow running dog forward, strong branching off from a short distance, hesitant, or getting a dog, the turbulent early postponement, possibly the early position of dog handler on arrival (changing positions) are assessed- points.

9. postponement of the long - 10 points

a) after one song goes: "lie", "sit".

b) implementation: before starting the exercises section B of another dog handler of the dog shall postpone the audible command "lie" to the location specified by him, not the referee let a guide or any other subject. The handler is going in exercise space, without looking back, at least 30 steps from the dog and remains in hiding outside its control. After a period of implementation exercises 1 to 7 other dog has dog, without influencing the handler, you lie. On the instruction of the handler goes to the dog and stands to his right side. After about 3 seconds, at the direction of the referee, gives the handler a dog sound command "sit", to which the dog must sit down quickly and directly.

c) rating: Restless behavior handler, as well as other assistance dog, his restless postponement, possibly the early position of dog handler on arrival (change location) is rated-points. Partial evaluation exercises it is possible, if the dog gets "up" or "sit" without leaving instead of postponing. If they leave the dog instead of the postponement of more than 3 m before the end of practice 5 another dog, the exercise is evaluated to 0. If they leave the dog instead of postponing the later will receive partial evaluation. If, when the arrival of the handler as the dog to him, followed by a loss of up to 3 points.

IPO 3 SECTION "C"

Exercise 1: search for dummy 10 points

Exercise 2: exposure and 10 points woofing

Exercise 3: attempt to escape the training dummy 10 points

Exercise 4: defense of the dog during guarding phase 20 points

Exercise 5: escorts from behind 5 points of training

Exercise 6: oversize dog when accompanied by 15 points

Exercise 7: attack on the dog from the movement of the 10 points

Exercise 8: defense of the dog when the babysitter 20 points

for a total of 100 points.

General provisions:

The longer sides of the suitable space is under construction 6 shelters (safe houses, 3 on each side-see diagram). The required markings must be for the handler of the arbitration and the dummy well visible. Helper shall be provided with protective clothing, a sleeve and a baton. The sleeve must be fitted with a protective layer which is the stitched jute natural colors. If it is necessary to constantly monitor the helper dog may become unconditionally without any movement. But it shall not make any threatening or defensive movements. Protective sleeve protects your body. It is left to the handler, the manner in which removes his knife dummy baton.

(see: "the general part – handler work"). In all stages of the trials may be working with one lesson in the defense. For all dog handlers competent training level tests must be used the same music. Dogs, which are not manageable, the desert inside the sleeve when the strong influence the (secession) handler, are disqualified, the evaluation does TSB. In dogs, who exercises the defense fail, or let them off, it is necessary to interrupt the assessment section C in section C, points stand, but perform guest TSB. An audio signal for releasing is allowed in all exercises of defense just once.

Table 1 evaluation for the kite (losing points)

Slow out	First additional command followed by immediate out	First additional command followed by slow out	Second additional command followed by immediate out	Second additional command followed by slow out	No out after second additional command, respectively additional influence
0,5 - 3,0	3,0	3,5 - 6,0	6,0	6,5 - 9,0	disqval.

1. locate the figurant - 5 points

a) after one song goes: "district", "with me" (audio signal "to me" can be associated with the name of the dog).

b) implementation: Helper is hidden away in the last hiding place so that the dog was not visible. The handler and his dog stand at the front of the first fences in order to send the dog to the six parties. Implementation of section (C) starts at the instruction of the referee. On a short sound signal, "district" and the command hold out right or left arm, which may be repeated, must run from the dog handler quickly and purposefully to the first refuge and this go around. Once the dog shelter goes, it's the handler calls an audible command "to me" (this may be associated with the name of the dog) to each other, and the movement it sends new sound command "district" to the next shelter. The handler for the search moves normal step after the imaginary central axis of the search space, you may leave. The dog must always be in front of the handler. The dog when it comes to hiding helper handler must stand, other audio and commands are not accepted.

c) rating: shortcomings in manageability, close and result-oriented circle the shelter are assessed-points.

2. exposure and bark 10 points

a) A sound command: "leg"

b) implementation: the dog must expose carefully and persistently helper woof. Not on the dummy jump, (touch) or bite. On the instructions of the referee, after about 20 seconds the barking, as the handler to a distance of 5 steps to behind the barking dog. To the next instruction of the dog's leg to withdraw and stand with him into the basic position.

c) rating: Shortcomings in continuous and woofing and look more convincing display (up to a separate sound command handler), which is not affected by the present arbitration or coming inside, are rated precipitation points. For intermittent barking is him given 5 points. If the dog barks faintly, are deducted 2 points. If a dog to bark, carefully and actively patrol the figurant is downed 5 points. When is the light touching dog into the dummy, shall be deducted up to 2 points, when you bite up to 9 points. If the dog leaves the helper before the referee will issue a handler instruction to leave the center of an imaginary axis, it can once again send a handler to his knife dummy. If the dog remains on the dummy can be in section (C) continued the exercise issue and woofing, but it is rated poorly. If the dog cannot be send again or leave the dummy again, assessment of section C is broken. With respect to the dog, opposite the handler or coming to come to him before the command, followed by partial evaluation in the context of the lack of signs.

3. attempt to escape the figurant 10 points

a) after one song goes: "stay", "Down", "let go". (b)) to perform: to instruct the referee asks the handler helper to leave the shelter. The place where you want to build, it is marked. On the instruction of the handler leaves the dog, then the designated place to its postponement. The distance between the lesson in defense and the dog must be 5 steps. The handler leaves the dog lying guard the dummy and goes to the shelter so that he could watch his dog out, dummy, and arbitration. On the instruction of the arbitration with the helper tries figurant to escape. Without hesitation, the dog must independently, energetic and strong bit to prevent escape. It may only take a bite into the protective sleeve of the figurant. At the direction of the music remains still, the dog must go immediately. For releasing separately can give the handler, at the appropriate time interval, one audio signal.

If you let the dog after the first audible signals, the referee handler instructed to issue up to two other audio commands to drop. If let the dog after the third audio signals (one and two additional), followed by disqualification. When giving a sound command of the "release" could have the handler without the dog interacted. After releasing the dog must remain close to the dummy to guard it closely.

c) rating: shortcomings in the implementation of important exercise requirements are evaluated precipitation points. In particular, the requirements are: a rapid and forceful response of the dog on the fading and his helper followed, culminating in a convincing retention with a full and peaceful bit to the drop, the attentive care just for dummy. If the dog remains to lie or to a distance of about 20 steps to escape detention, and bit does not prevent the figurant is the assessment of section C terminated. If the dog is slightly distracted when the babysitter or his work slightly impure, it is reduced to about one practice assessment mark. If the dog was guarding the figurant or waving his very work unclean, it is reduced to two practice assessment mark. If access to helper, dog though he remains a guest is reduced by three marks. If a dog leaves the figurant or receives the audio signal from the handler to the dummy, he is the guest of section C terminated.

4. defense of the dog during guarding phase - 20 points

a) after one song goes: "let go", "stay".

b) implementation: after the babysitter helper dog, which takes about 5 seconds, on the instruction of music will take the attack on the dog. Without influencing the dog handler must be energetic and strong bit help. The dog may bite when it only into the protective sleeve of the figurant. When the dog bites, wounds 2, gets a baton. The wounds are allowed only on the thighs or the rear of the dog. On the instructions of the referee will then calmly State, the helper dog it must be immediately let go. For the kite can put the handler in the interval separately one audio signal.

If you let the dog after the first permitted by command, you can judge the handler to issue up to two other audio commands "drop". If you let the dog after the third audio signals (1 and 2 additional), followed by disqualification. At the time of the command "drop" it must calmly State the handler without the dog interacted. After releasing the dog must remain in the dummy, and guard it carefully. On the instruction of the handler as a normal step directly toward the dog and an audible command "stay" is the basic attitude with him. The Club does not remove his knife dummy.

c) rating: shortcomings in the implementation of important exercise requirements are evaluated precipitation points. In particular, the requirements are: quick and compelling detention, full and calm until the bit releasing, attentive care just for dummy. If the dog is slightly distracted when the babysitter or his work slightly impure, it is reduced to about one practice assessment mark. If the dog was guarding the figurant or waving his very work unclean, it is reduced to two practice assessment mark. If access to the helper dog even though, remains a guest is reduced by three marks. When it comes to dog opposite coming handler, the exercise is rated poorly. If a dog leaves the dummy before instructing the arbitration to the arrival of the handler or the handler receives the audio signal from the dummy, he is the guest of section C terminated.

5. escort from behind by 5 points

a) 1 audio command: "leg"

b) implementation: following the exercise no. 4 followed by an escort back to the dummy a distance of about 30 feet. The progress of the escorts shall be determined by arbitration. Handler prompts the walking dummy and goes with the dog, which he has freely at his feet, at a distance of 5 steps behind him. The dog must carefully guard the helper. The distance of the 5 steps must be adhered to in the course of the entire entourage.

c) rating: shortcomings in the implementation of important exercise requirements are evaluated precipitation points. In particular, the requirements are: an attentive monitoring of the figurant, a quiet walk the dog, with a distance of 5 steps.

6. spillway dog when accompanied and one audio command by 15 points

a) "let go", "stay".

b) implementation: When accompanied by the referee, without stopping the spillway dog lesson in defense. Without the influence of the handler must be a dog without hesitation, energetic and strong bit help. The dog may bite while only into the protective sleeve of the figurant. When the dog bites, the handler must remain on the site you are currently located. On the instructions of the referee will then calmly State, the helper dog it must be immediately let go. For the kite can put the handler in the interval separately one audio signal. If you let the dog after the first permitted by command, you can judge the audio handler to issue up to two other audio commands "drop".

If let the dog after the third audio signals (one and two additional), followed by disqualification. At the time of the command "drop" it must calmly State the handler without the dog interacted. After releasing the dog must remain in the dummy, and carefully guard it. On the instruction of the handler as a normal step directly toward the dog and an audible command "stay" is the basic attitude with him. Baton takes his knife dummy, followed by the side accompaniment to arbitration on the dummy a distance of about 20 steps. One audio signal "to the leg" is enabled. The dog goes after the right side of the dummy, so it is between lesson in defense and the handler. For escorts must closely guard the helper dog may not jumped on him, nor can it invade. Before the arbitration, this group of stops, the dog sits down, the handler passes the baton to arbitration and he will produce their first part of the section C.

c) rating: shortcomings in the implementation of important exercise requirements are evaluated precipitation points. In particular, the requirements are: quick and compelling detention, full and calm until the bit releasing, attentive care just for dummy. If the dog is slightly distracted when the babysitter or his work slightly impure, it is reduced to about one practice assessment mark. If the dog was guarding the figurant or waving his very work unclean, it is reduced to two practice assessment mark. If access, to the helper dog even though remains a guest is reduced by three marks. When it comes to dog opposite coming handler, the exercise is rated poorly. If a dog leaves the dummy before instructing the arbitration to the arrival of the handler or the handler receives the audio signal from the dummy, he is the guest of section C terminated.

7. attack on the dog and the movement of the 10 points

a) after one song goes: "sit", "stay", "let go".

b) implementation: the dog handler stands on the marked place in the level of the first shelter on the imaginary central axis of the space. A dog can hold hand on the collar, but it must not interfere with. On the instructions of the referee will perform music, equipped with a baton from the hiding place and runs to the central axis of the space. Once this reaches, without interrupting the run, turns frontally against the handler and the dog and screaming moves on them run attack. Once closer to the handler and dog, at a distance of about 60 steps, dog handler to instruct the referee sends to defend an audible command "keep". The dog must attack the energetic and strong to prevent bit. The bit must be conducted only into the protective sleeve of the figurant. The handler must not leave the station. On the instruction of the arbitration stay then music, calmly State, it must be immediately let go of the dog. For releasing the handler can give individually in the appropriate time interval, by one sound signal.

If handler let the dog after the first permitted by command, you can judge the audio handler to issue up to two other audio commands "drop". If let the dog after the third audio signals (one and two additional), followed by disqualification. At the time of the command "drop" it must calmly State the handler without the dog interacted. After releasing the dog must remain in the dummy, and carefully guard it.

c) rating: shortcomings in the implementation of important exercise requirements are evaluated precipitation points. In particular, the requirements are: quick and compelling detention, full and calm until the bit releasing, attentive care just for dummy. If the dog is slightly distracted when the babysitter or his work slightly impure, it is reduced to about one practice assessment mark. If the dog was guarding the figurant or waving his very work unclean, it is reduced to two practice assessment mark. If access, to the helper dog even though remains a guest is reduced by three marks. If a dog leaves the figurant or receives the audio signal from the handler to the dummy, he is the guest of section C terminated.

8. dog defense when guarding the 20 points

a) one audio command "let go", "stay" (2 x).

b) implementation: after the babysitter helper dog, which takes about 5 seconds, the referee shall take the music, the attack on the dog. Without influencing the dog handler must be energetic and strong bit help. The dog may bite when it only into the protective sleeve of the figurant. When the dog bites, wounds 2, gets a baton. The wounds are allowed only on the thighs or the rear of the dog. On the instructions of the referee will then calmly State, the helper dog it must be immediately let go. For the kite can put the handler in the interval separately one audio signal. If you let the dog after the first permitted by command, you can judge the audio handler to issue up to two other audio commands "drop". If let the dog after the third audio signals (one and two additional), followed by disqualification.

At the time of the command "drop" it must calmly State the handler without the dog interacted. After releasing the dog must remain in the dummy, and carefully guard it. On the instruction of the handler as a normal step directly toward dog and an audible command "stay" is the basic attitude with him. Baton takes his knife dummy. It is followed by the side accompaniment to arbitration on the dummy a distance of about 20 steps. One audio signal "to the leg" is enabled. The dog goes after the right side of the dummy, so is located between lesson in defense and the handler. For escorts must closely guard the helper dog may not jumped on him, nor can it invade. Before the arbitration, this group of stops, the dog sits down, the handler passes the baton to an arbitration and reports him to their section C. Prior evaluation of the handler's on the instruction of the arbitration shall be a dog on a leash.

c) rating: shortcomings in the implementation of important exercise requirements are evaluated precipitation points. In particular, the requirements are: quick and compelling detention, full and calm until the bit releasing, attentive care just for dummy. If the dog is slightly distracted when the babysitter or his work slightly impure, it is reduced to about one practice assessment mark. If the dog was guarding the figurant or waving his very work unclean, it is reduced to two practice assessment mark. If access, to the helper dog even though remains a guest is reduced by three marks. When it comes to dog opposite coming handler, the exercise is rated poorly. If a dog leaves the dummy before instructing the arbitration to the arrival of the handler, or receives the audio signal from the handler to the dummy, he is the guest of section C terminated.

2.4 Selection team

At the Championship of the Czech Republic is necessary to meet the previous year exam IPO 3 with a minimum of 270 points. At the end of the previous year in the Advisory list of places where the selection races will be held. A contestant chooses 2 of them and sends the application, together with a copy of the test compound. The handler may take a maximum of 2 selection races, and if he wants to take part in the FCI World Cup/Championships, it is necessary to participate in the 2 selection races and meet the minimum limit, i.e.: 210 points. Points are calculated only from 1 (better) and from these points determine the ranking of the Championship of the Czech Republic. FCI World Cup/Championships counts better select a race and the number of points from the Championship of the Czech Republic. FCI World Championships each year qualify 5 athletes (Daušová, 2009). Rating test IPO presented table 2.

Table 2. Ranking tests IPO

Highest score	Excelent	Very good	Good	Satisfactory	Disappointingly
300	300- 286	285 - 270	269 - 240	239 - 210	209 - 0
100	100 - 96	95,5 - 90	89,5 - 80	79,5 - 70	69,5 - 0

3. Material and Methods

From the result of the World Cup/Championship FCI and the Championship of the Czech Republic, has evaluated the difference in success between two breeds most enjoyed ones currently in service, but also sports. This is an evaluation of the results of the contests, points that dogs of two investigational breeds have achieved. Is not taken into account environmental conditions (handler, arbitration, venue and others), the guest is on without taking into account the gender and age of animals and other aspects. It is only a comparison of score results of two breeds, the most used dogs in business and sports on IPO World Championships (World Cup FCI) and the Czech Republic IPO Championship (Czech IPO, CZ IPO) (IPO - International trial rules) for several years. To determine the actual quality and usability of the dogs would have to be taken into account and, above all to create such conditions that would allow the implementation of testing selected dogs with the selected handler, the arbitrator and the same terms of environment.

From these results, the number of dogs competing at competitions it is possible to provide for the development of popularity or enlargement of the individual breeds. These numbers are applied in the methodology for presenting not only the total number of dogs of both breeds for all reference years also, but also the differences in numbers and representation for individual years. The description of the individual disciplines (no copy from web) - defense, obedience, track - presented in the literature review. It evaluates results from IPO for breeds German Shepherd (GS) and Belgian Shepherd Malinois (BSM).

Table No 3 below (in the chapter results) presents the number of dogs of the breed German Shepherd and Belgian Shepherd Malinois which participated in the competition for each of the years during the World Championship FCI.

For the evaluation of the results from the World Championship FCI (WC IPO FCI) were total 799 dogs in the years 2003 - 2011. German Shepherd dogs were - 418, Belgian Shepherd Malinois - 81.

Table 4 below (in the chapter results) presents the number of dogs of the breed German Shepherd and Belgian Shepherd Malinois, who participated in the competition for each of the years in the Championship of the Czech Republic IPO. For the evaluation of the results of the Championship was for the years 2003 IPO to 2011 rated overall 344 dogs, German Shepherd were 238 and 106 were Belgian Shepherd Malinois. Compared with the performances (in disciplines such as obedience, defense and track + score overall) of both breeds in the years 2003 – 2011 and average at these competitions.

The statistical evaluation of the program was used to calculate the SAS 9.2. Implemented the evaluation for years (the differences between breeds for the individual years 2003 and 2011) and also for all the years together in both breeds. For the first comparison of the average values obtained were used in each of the disciplines and points total, so the total score. Evaluate the hypothesis: "Belgian Shepherd Malinois (BSM) will have higher scores than German Shepherd (GS)". For statistical hypothesis testing was needed to determine whether the data files are in the normal distribution. To determine whether the file has a normal distribution is used "to Test the good matches." Because the Test of goodness of fit demonstrated that the data do not have a normal distribution, has chosen for the calculation of the nonparametric test for testing hypotheses. For this reason, for further testing was elected the "Wilcoxon two-sample test". Wilcoxon two-sample test is nonparametric similar to the t-test for independent files. Marked bold are statistically significantly different values.

4. Results

As assessed by two independent files, FCI World Cup/Championship IPO and Championship of the Czech Republic (CZ IPO), the results will be presented separately. At the same time for each contest, and each year the number of wins on behalf of dogs of both breeds.

4.1 Representation of individual breeds

The first evaluation was to determine the representation of individual races in the competition. From these results you can provide for the development of popularity, or enlargement of the individual breeds.

Table 3. Number of dogs WC FCI IPO

Year/Breed	2003	2005	2006	2007	2008	2009	2010	2011
BSM	42	45	54	43	45	61	44	47
GS	50	52	34	54	62	59	56	51

In 2004, MS FCI has not held. On MS FCI IPO are two breeds represented in balanced numbers, therefore, it is possible to say that there is no preferred neither of investigational breeds. The representation of the breed has a rather unstable character.

Table 4. Number of dogs CZ IPO

Breed/year	2003	2004	2005	2006	2007	2008	2009	2010	2011
BSM	3	5	11	11	12	19	13	22	10
GS	33	32	32	27	28	23	27	19	17

As regards the Championship CZ IPO in table 4 it is clear that increased the number of Belgian shepherd Malinois since 2003, who attended the contest at the top of the IPO. In recent years German Shepherd was seen a slight decrease in attendance. In Czech Republic increase the popularity of the breed Belgian Shepherd Malinois.

4.2. Comparison of score (breeds and disciplines) WC FCI IPO and CZ IPO

The basis for evaluation of the statistical hypothesis testing is to determine the performance difference between the two breeds of the diameters obtained points, as in the overall ranking, as well as in individual disciplines. First results are presented from WC FCI IPO.

4.2.1 Comparison of score (both breeds and disciplines) WC FCI IPO

Table 5. Average total score WC FCI IPO

Breed/year	average	2003	2005	2006	2007	2008	2009	2010	2011
BSM	262	272,1	268,4	260,9	250,5	259,7	254,7	259,2	266,8
GS	250	260,2	249,3	250,8	236,6	249,1	241,0	235,9	259,9

Table 5 and Graph 1, it is evident that in the total points are the total results of the Belgian shepherds better than German shepherds and about 12 points, which is about 5 %, this difference is statistically inconclusive. Furthermore, the results of individual years, it is clear that the Belgian Shepherd was always higher in the MS FCI overall assessment. Statistically conclusive difference is in the years 2003, 2005, 2007, 2008, 2010, and 2011. These statistically significant values in the table are marked bold. Graph 1 presents the spot not only results, but also the development and differences in the individual breeds between the years.

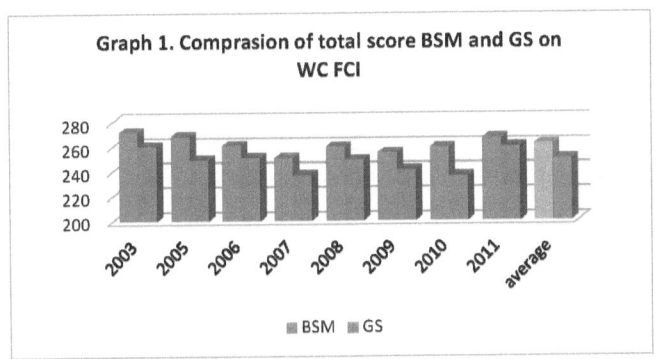

Tab 6a, b, c, Average values for the track, obedience and defense for all years from WC FCI IPO between BSM and GS

Tab 6a Track Breed/year	2003	2005	2006	2007	2008	2009	2010	2011	average
BSM	95,8	88,4	82,2	72	79,6	86,6	87,3	91,2	85,4
GS	92,8	84	76	67,5	76,4	83,6	73,7	92,4	80,7
Tab 6b									
Obedience	2003	2005	2006	2007	2008	2009	2010	2011	average
BSM	87,5	90,2	88,8	89,3	88,1	86,5	87,1	86,9	87,9
GS	84,1	83,4	87,3	86,4	84,7	83,3	83,9	83,2	84,5

Tab 6c									
Defense	2003	2005	2006	2007	2008	2009	2010	2011	average
BSM	88,8	89,7	89,9	89,2	92,1	87,4	84,9	88,7	88,8
GS	83,2	81,9	87,5	82,6	88	83	78,3	84,3	83,5

Table 6 describes the average value for each discipline for all reference years on the WC FCI. In each of the disciplines, the situation is different as compared to the overall results. In table 6a and you can see that in addition to the year 2011 is a Belgian Shepherd better in the discipline of track than a German Shepherd. In 2011, the German Shepherd was better in the average guest. The differences are statistically inconclusive, in addition to the year 2003. Table 6b shows the average values for obedience. The Belgian Shepherd Malinois is in each year and in the overall average better, and in addition to the year 2006, the results are statistically significant. Table 6 shows that in all the years is Belgian Shepherd in defense better and besides 2006 even statistically relevant. The results for each of the years presented table 6 a,b,c and table 7 (average for the reference years), and at the same time for clarity and demonstration of development point rating of individual breeds and the differences between them are presented in graphs 2, 3, 4.

Table 7 Average values of the individual disciplines for all years WC FCI IPO

Breed	Disciplines - average		
	Track	Obedience	Defense
BSM	85,4	87,9	88,8
GS	80,7	84,5	83,5

The overall average value of individual disciplines for all reference years on the WC FCI IPO shows table 7 and Belgian Shepherd Malinois received higher score in all disciplines. In the trail of statistically we found no significance in obedience and defense the statistically relevant.

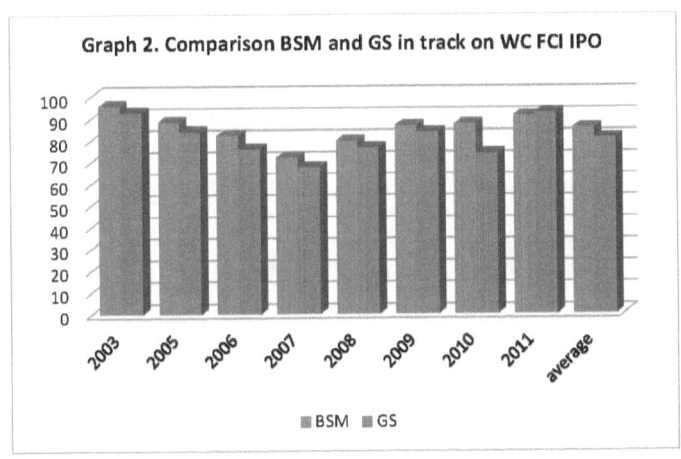

Graph 2. Comparison BSM and GS in track on WC FCI IPO

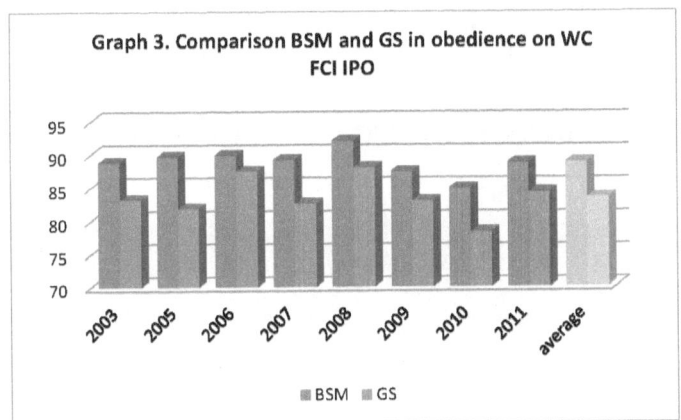

Graph 3. Comparison BSM and GS in obedience on WC FCI IPO

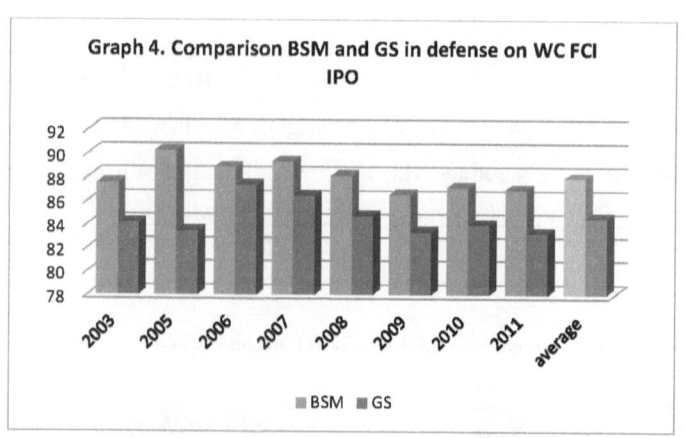

Graph 4. Comparison BSM and GS in defense on WC FCI IPO

4.2.2 Comparison of score (both breeds and disciplines) CZ IPO

Tab. 8 Average value of the total results CZ IPO

Breed	average	2003	2004	2005	2006	2007	2008	2009	2010	2011
BSM	253	253	271	**264**	251	249	229	268	262	271
GS	249	234	250	**249**	252	255	247	260	257	267

The overall results on CZ IPO in Czech Republic are statistically significant for the average rating for the reference period and also in 2005. In each of the years, you can monitor the performance differences between breeds. In the years 2003, 2004, 2005, 2009, 2010 and 2011 received higher scores Belgian Shepherd Malinois in the years 2006, 2007 and 2008 received higher scores German Shepherd. Marked bold are statistically significantly different values.

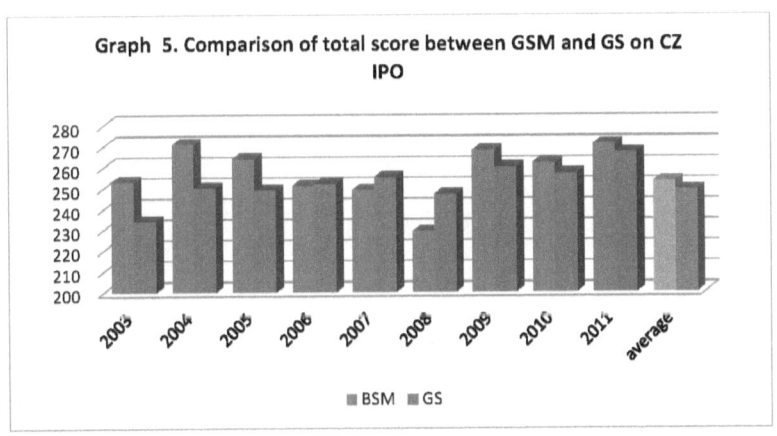

It was found that the competition in the Czech Republic also Belgian Shepherd Malinois (in addition to the years 2006, 2007 and 2008) received higher scores overall than a German Shepherd. In graph 5 is a noticeable development for individual breeds and the difference between them.

Table 9a, b, c The average score for the track, obedience and defense between breeds - CZ IPO

Tab 9a Track	2003	2004	2005	2006	2007	2008	2009	2010	2011	average
BSM	78,7	88,6	89,4	72,6	72,8	**56,9**	89,2	83,8	90,5	**78,5**
GS	78,8	80,3	82,2	82,7	82,6	**72,7**	83,7	81,2	93,8	**81,6**
Tab 9b										
Obedience	2003	2004	2005	2006	2007	2008	2009	2010	2011	average
BSM	86,3	92,8	85,2	87,5	86,1	87,2	85,1	81,6	89,7	**85,9**
GS	83	86,4	80,4	82,6	84,8	85,3	84,9	82,7	85,4	**83,9**

Tab 9c										
Defense	2003	2004	2005	2006	2007	2008	2009	2010	2011	average
BSM	87,7	89,8	89,2	90,9	90,2	84,8	**93,8**	88,5	90,5	**89,2**
GS	72,6	82,8	85,9	87	87,1	88,7	77,4	88,7	88	**83,6**

Table 9 a shows that in the discipline of track received higher scores in years 2004, 2005, 2009 and 2010 Belgian Shepherd Malinois and in 2003, 2006, 2007, 2008 and 2011 received higher scores German Shepherd. Statistically significant results (bold marked) are in 2008 and an average for the reference years. The German Shepherd received higher scores in track of CZ IPO than a Belgian Shepherd. Table 9b shows that in addition to the year 2010, when it was a German Shepherd, Belgian Shepherd received higher scores (higher ratings) in obedience. Statistically significant are the averages of obedience for all years. In table 9c you can see that in addition to the years 2008 and 2010 that a Belgian Shepherd received higher scores in defense, with a statistically significant difference is only in the year 2009 and in the overall average for all years. Marked bold are statistically significant differences.

Table 10. Average values together CZ IPO

Breed	Disciplines - average		
	Track	Obedience	Defense
BSM	**78,5**	**85,9**	**89,2**
GS	**81,6**	**83,9**	**83,6**

Table 10 shows the average values of the individual disciplines during the reporting period and all of them are all statistically significant (bold marked). In the discipline of track received higher scores German Shepherd, in disciplines obedience and defense received higher scores of the Belgian Shepherd. All the results are statistically significant. From the tables 9a, b, c and table 10 (average for the reference years) are seen better scores (results) in track of German Shepherd and Belgian Shepherd Malinois received higher scores in obedience and defense. At the same time for clarity and demonstration of development point rating of individual breeds and the differences between them are presented in graphs 6, 7, 8.

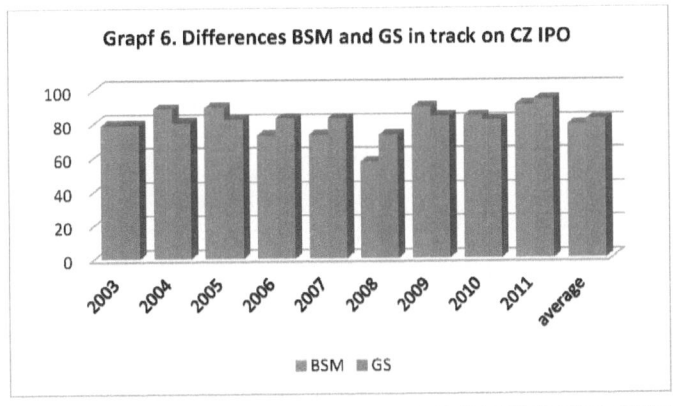

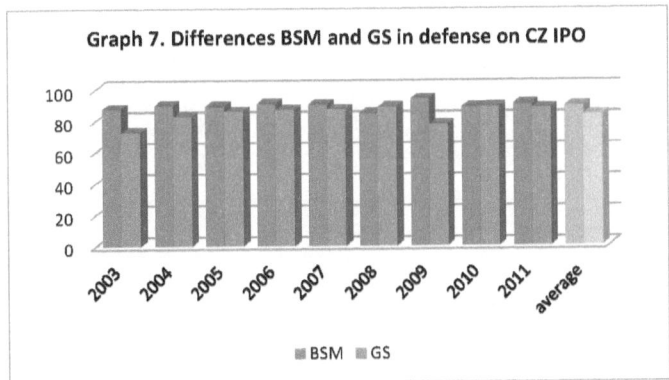

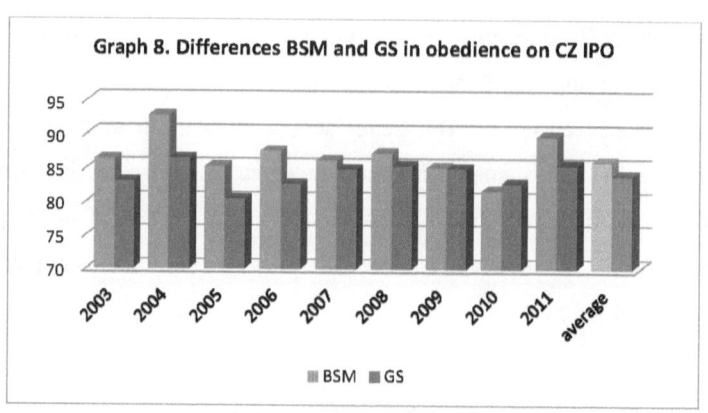

Table 11. The values of the test criteria normal approximations WC IPO

	Total	Track	Obedience	Defense
total	**7,01**	1,2	**7,24**	**7,43**
2003	**3,05**	**2,03**	**2,34**	**2,20**
2005	**3,18**	1,13	**4,17**	**3,64**
2006	1,78	0,57	1,80	1,39
2007	**2,05**	0,02	**1,97**	**3,46**
2008	**2,52**	0,87	**2,93**	**3,74**
2009	1,47	1,01	**2,55**	**2,19**
2010	**3,20**	**1,81**	**2,94**	**2,18**
2011	**2,75**	0,25	**2,32**	**3,16**

In the tables 11 and 12 are red marked values when saved for α>uαkrit when the differences are statistically significant mean values observed. WC FCI in the years 2006 and 2009 in the overall ranking was not a statistically significant difference. Overall, it is better (received higher scores) Belgian Shepherd and in the average rating in each of the disciplines, in track received higher scores in addition to the 2011 than German Shepherd. Statistically conclusive difference is only in the year 2003. In obedience and defense received higher scores Belgian Shepherd always, and in addition to the year 2006 even statistically relevant.

Table 12. The values of the test criteria normal approximations CZ IPO

	Total	Track	Obedience	Defense
total	**2,01**	**2,18**	**3,07**	**3,62**
2003	1,12	0,66	0,31	0,49
2004	1,91	0,31	1,78	0,53
2005	**2,28**	0,31	1,64	1,36
2006	0,82	1,69	1,37	0,60
2007	0,24	1,33	0,21	1,56
2008	1,88	**2,31**	1,28	0,09
2009	1,40	0,94	0,13	**3,65**
2010	0,26	0,42	1,20	0,68
2011	0,80	1,71	1,56	0,80

The overall results are statistically significant and also in 2005. Belgian Shepherd Malinois received higher scores in the years 2003, 2005, 2009, 2010 and 2011, the German Shepherd received higher scores in the years 2006, 2007, and 2008.

In the track of the Belgian Shepherd received higher scores in the years 2004, 2005, 2009 and 2010, in 2003, 2006, 2007, 2008 and 2011 received higher scores German Shepherd. Statistically conclusive difference is in 2008, and in the overall average. In obedience with the exception of the year 2010 received higher scores Belgian Shepherd Malinois. Statistically conclusive difference is only in the overall ranking of obedience. In defense of in addition to the years 2008 and 2010 also Belgian shepherds achieve better results. A statistically significant difference is in 2009 and in the overall assessment of the defense.

5. Discussion

Whereas a similar evaluation was nowhere so far carried out, it is quite difficult to assess and discuss our results with other top foreign materials. Regarding the breed German Shepherd is used for a number of years in service and sports purposes. It is the most popular breed of dog worldwide and the most widely used breed (Slámová, 2011). Their intelligence, grace lines and other valuable properties fidelity and watchfulness especially, became favorite breed for the breeders around the world (Najmanová and Humpál, 1981).

Originally was bred as a shepherd dog. One of the first roles was the role of a German Shepherd as police dog. Stephanitz induced the German Government that the German Shepherd dog recognized as suitable for police work. It was the beginning of the recovery of the breed as the military and service dog. In 1903 the SV recommended police authorities take advantage of German Shepherd dogs and 1903 started tests of their ability. The results have been satisfactory enough that the police organizations in many big cities, was recommended that dogs try and where appropriate accept them as an integral part of the system of law enforcement (Kranátová, 2001).

It is currently the most used worldwide German Shepherd dog for working purposes. From the armed forces, through rescue, Braille and assistance dogs after all dog sports, where achieves great results (Trávníčková, 2005).

And in the Czech Republic is currently the German Shepherd breed in 2008 also in the security forces. This is a breed with a versatile use up approximately 90 % from all breeds, is considered to be the most easily trainable breed. Thanks to its characteristics has become the most successful service dog in the world. German Shepherd dogs are among the most adaptable, versatile and successful breeds (Kranátová, 2001). On the contrary, according to the Boss (1989) in the United States of America are used about 50 % Belgian shepherds and 37 % German shepherds as professional dog in armed forces. A higher temperament and stamina of Belgian Shepherd Malinois leads for the US military to increase over German shepherds, which were until then the preferred breed (Burghardt, 2003).

At competitions achieves German Shepherd very good and with stabile results, which are evident even from the appreciation in this work. 1980 in Central Europe experienced a remarkable flowering of Belgian Shepherd Malinois. Although in the past he knew only a few real lovers of Belgian shepherds, it suddenly grew extremely popular dog used in sports, which aroused positive interest also for the police and customs authorities. Malinois are for example in recent years occupying good position at the World Championship FCI IPO (Vacková, 2008). Belgian Shepherd dog has 4 varieties, but for the work it is best to use Malinois (Bossi, 1989). For example in the USA is in the sports and business a clear trend away from German shepherds to Malinois, according to reports and the presented information intended for foreign sending this breed is characterized by a greater taste for work than a German Shepherd (Malik, 2012).

It is evident the higher use of Belgian shepherds Malinois than German shepherds. In addition to the health for all breeds Belgian Shepherd plays an important role also for greater agility, speed and less weight. Also the plethora of greed which can sometimes be on the injury, but that the German shepherds often missing (Schiller, 2012). Police in North Rhine-Westphalia recedes from German shepherds using as service dogs, replaced Malinois. This is in addition to health and high prices also a small resistance of German shepherds (Bonke, 2012). By contrast the opinion of the head of the Central Police Department of Cynology by Richard Haas more appropriate for business use is the breed German Shepherd. It is for reasons of easier adaptability to change handler, better management of the bulk of the accommodation in the official kennels and higher concentration at work, unlike the Belgian Shepherd Dog, which is a very hectic, poorly manages changes to a canine unit and he's not kidnapping well bulk of housing (Haas, 2012, pers. comm.).

Also the current Central Breeding Advisor judge the exterior of German shepherds, participant selection races Jiří Novotný thinks that German Shepherd is a versatile and reliable breed in particular, where the dog requires more demanding work, must work for a longer period and is also in many cases forced to improvise, i.e. "the thinking behind the handler," must deal with demanding situations where service rest takes place under nonstandard conditions. The Belgian Shepherd Malinois is successfully deployed, where there are performances of the simple, action and do not last long (Novotný, 2012, pers. comm.).

Both breeds have a high popularity in the historic representation of underserved working breeds worldwide. This is due to their character, reliability, training and intelligence. German Shepherd is one of the big dogs, Belgian Shepherd dog is a medium sized dog. Belgian Shepherd is faster in a speech and has a better temperament, German Shepherd is less demanding on housing and treatment than Belgian Shepherd (Wedding, 2004). Working dogs are used for different purposes in the police force to patrol, detection of narcotics and explosives. In the current political climate, there is a growing demand for these specially trained dogs and this is related to the research in this area. The results show that for some of the tasks are certain breeds better, sometimes plays role the difference and gender (Slabbert and Odendaal, 1999). It is further confirmed that the behavior of individual dogs, regardless of the gender affects right education and environment (Svatberg, 2005).

Just important is correct rearing, then new owner for each dog (Burghardt, 2003). The results of the work showed that on WC IPO FCI has improved the overall results of Belgian Shepherd Malinois in most reporting years statistically relevant.

In CZ IPO received higher scores Belgian Shepherd dogs in the overall averages in the years 2003, 2004, 2005, 2009, 2010 and 2011, the German Shepherd received higher scores in the years 2006, 2007 and 2008. Statistically significant are for the total period and in 2005. In the discipline of track received higher scores overall better German Shepherd and even statistically relevant. Belgian Shepherd received higher scores in track 2004, 2005, 2009 and 2010, otherwise in 2003, 2006, 2007, 2008 and 2011 received higher scores German Shepherd. In obedience received higher scores Belgian Shepherd in addition to the 2010 statistically nonsignificant, an overall assessment of obedience clearly. In defense, in addition to the year 2010 the 2008 Belgian Shepherd received higher scores, statistically significant results were in 2009, and the overall averages. Koráb (2008) got results that Belgian Shepherd received higher scores on the World Cup in track, such as on the Championship and the German shepherd dog received higher scores in obedience and defense.

An experienced participant of selection races, Championship and the World Cup, William Balej, who competed with the Belgian and German shepherds, says: "Belgian Shepherd is preferable to breed on the competition in IPO than a German Shepherd, is more active in training, harder and faster" (Balej, 2012, pers.com.) This unique point of view, more or less confirms Jiří Novotný: "German Shepherd as a breed to use maple, like a dog for the protection of the family, sheep dog, use in the armed forces, use for demanding scent work only after the dog for the so-called top sport. This corresponds to his physique, his nervous system and his wide variability. Belgian Shepherd was bred by contrast mostly as a dog for the top sport, this corresponds to the light construction of the body, the ability to develop a high speed, at the same time and perseverance, especially in obedience".

At competitions, according to the IPO is a Belgian Shepherd very impressively especially when performing disciplines of obedience and defense. Most of the Belgian shepherds plays to counterattack the figurant with dedication and jumping against him from a great distance. If however comes at the hitchhiking and the tracks are laying in the worst terrain and to their development occurs around noon most of the Belgian shepherds in this discipline fails. Thanks to this moment are more or less forces at competitions between the German and Belgian shepherds settled (Novotný, 2012, pers. comm.).

Overall averages may in certain years on both contests affect the greater amount of zero ratings that do not match the quality of the individual breeds, but can be caused by an error handler, weather conditions, and the legendary "racing happiness".

6. Conclusion

The aim of this work was to summarize the history and present of the Belgian Shepherd Malinois and German Shepherd, present their conditions for breeding to purebred breeding, use of both breeds and working mainly in sports training from the IPO and the success of both breeds in the competitions. These two breeds are most commonly used for sports and business training. Evaluate the differences of both breeds in Czech IPO Championship (CZ IPO Championship) and WC (World Cup/Championship) IPO. The comparison was monitored among 2003 - 2011 and for the nine year period and in each of the disciplines.

It was found that at the World Cup in the overall ranking for all reference years received higher scores Belgian Shepherd Malinois than German Shepherd. The results for each of the years, it is clear that the Belgian Shepherd always received higher scores. In addition to the years 2006 and 2009 and results are statistically significant. In track in addition to the 2011 received higher scores Belgian Shepherd in the overall average for the reference period. Statistically conclusive difference was only in the year 2003. In disciplines obedience and defense received higher scores always Belgian Shepherd and even the overall averages for all years. In addition to the year 2006, the results are statistically significant even in the overall averages.

Belgian Shepherd Malinois received higher scores on CZ IPO in the overall evaluation of all disciplines. The results were statistically significant and also in 2005. In each of the years were performance differences between breeds. Belgian Shepherd received higher scores 2003, 2004, 2005, 2009, 2010 and 2011 and German Shepherd n years 2006, 2007 and 2008. In track German Shepherd in the overall diameter received higher scores and even statistically relevant. In each of the years the Belgian Shepherd received higher scores in 2004, 2005, 2009 and 2010 and German Shepherd in 2003, 2006, 2007, 2008 and 2011. In 2005, it is also statistically conclusive difference. In obedience, in addition to the year 2010, received higher scores Belgian Shepherd. In each of the years, the results are inconclusive. The overall averages for obedience are statistically significant. In defense, in addition to the year 2008, and 2010 received higher scores Belgian Shepherd, with a statistically significant difference is only in the year 2009, and also in the overall assessment of the defense.

It is evident that the original hypothesis "Belgian Shepherd Malinois is achieved in the competitive disciplines, according to the IPO better than German Shepherd", was confirmed, although, of course, the evaluation of contests results - points that dogs of two breeds achieved.

7. Literature

Allan, R., Allanová,C. 1997. Německý ovčák. nakl. TIMY s.r.o. Bratislava, str.115, ISBN 80-88799-44-9.

Bonke, G. 2012. NRW- Polizei muster Deutche Schaferhunde aus. [cit. 2012-4-2] dostupné z <http://www.spiegel.de/panorama/0,1518,781643,00.html>

Balej, V. 2012. pers.comm. Bělá p.B. 2012

Bossi, E. 1989. The Belgian shepherd dog and his history. Ifolith fotolizho Ges.m.b.H. And Co., KG.

Burghardt, W. 2003. Behavioral considerations in the management of working dogs. Veterinary Clinical Small Animals. 33:(417 - 446).

Císařovský, M. 2011. Rytmistr von Stephanitz a jeho německý ovčák. Pes přítel člověka 5, Pražská vydavatelská společnost. p. 20 - 21. ISSN 0231-5424.

Courreau J.F., Langlois B. 2005. Genetic parameters and environmental effects which characterise the defence ability of the Belgian shepherd dog. Applied Animal Behaviour Science. 91:(233 – 245).

Daušová, Z. 2009. pers. comm. Praha 2009

Divišová, K., Podešťová, M., Benda, J. 2003. První krůčky agility, nakl. PLOT, p.25, ISBN 80-86523-26-8

Fogle, B. 1996. Německý ovčák. nakl.ART AREA s.r.o.Bratislava. ISBN 80-967 573-2-6.

Haas, R. 2012. pers. comm. Domašín 2012

Koráb, S. 2008. Chov belgického ovčáka malinois se zaměřením na využití ve sportovní kynologii dle Mezinárodního zkušebního řádu IPO. Bakalářská práce. Česká zemědělská univerzita v Praze. Fakulta agrobiologie, potravinových a přírodních zdrojů. Praha. 41s.

Malik, W.2012. Belgischer Schaferhunde. [cit. 2012-2-20] available http://www.yorkie-hundeforum.com/forum/hunderassen/13/258648000.html.

Malíková, L. 2008. Klub tance se psem [cit. 2008-10-6], available < www.tanecsepsem.cz.>

Matušková, S. 1998. Belgičtí ovčáci. nakl. DONA České Budějovice.str.18-19, ISBN 80-85463-98-9.

Metzová, G. 2007. Německý ovčák. Svět psů. Minerva Praha. 9:(15–18). ISSN 1211-2976.

Najmanová, D., Humpál, Z. 1981. Atlas plemen psů. SZN Praha. p.62, ISBN 07-126-81.

Novák, K. 2008. Chovnost NO. [cit. 2008-10-6], available < www.ovčouni.cz>

Novotný, J. 2012. pers. comm. Mělník 2012

Pisarčíková, H. 1997. Historie chovu belgického ovčáka v České republice. [cit. 2008-10-5] available < www.zkovarny.com.>

Pisarčíková, H. 2006. Historie belgického ovčáka, [cit. 2008-10-6], available <www.kchbo.com.>

Schiller, P. 2012 Belgischer schaferhund. [cit. 2012-3-2], available < http://malinois-forum.de/forum/archive/index.php/t-35420.html

Sedlák, J. 2008. Svaz záchranných brigád kynologů [cit 2008-10-6], available < www.zachranari.cz.>

Slabbert, J.M., Odendaal, J.S. 1999. Early prediction of adult police dog efficiency-a longitudinal study. Applied Animal Behaviour Science 64:(269-288).

Soukupová, E. 2006. Německý ovčák. nakl. PLOT. 93, p.16, ISBN 80-86523-70-5.

Sinn D.L., Gosund S.D., Hilliard S. 2010. Personality and performance in military working dogs. Applied Animal Behaviour Science. 127: (51-65).

Stibůrek, J. 2002. Nejznámější ovčák. Pes přítel člověka. Pražská vydavatelská společnost. 11:(14–18). ISSN 0231-5424.

Svartberg, K. 2005. A comparison of behaviour in test and in everyday life of three lon sistent boldnes srelatep personality trits in dogs. Applied Animal Behavioral Science. 91:(103-128).

Šiška, J., Jánský, L. 2006. Nejrozšířenější plemeno na světě – německý ovčák, Pes přítel člověka. Pražská vydavatelská společnost. 3:(18–20). ISSN 0231-5424.

Šveráková, L., Hodek, J. 2012. Zkušební řád pro mezinárodní zkoušky pracovních psů a mezinárodní zkoušku psa-stopaře. Vydal ČKS Praha, p. 67–68.

Trávníčková, M. 2005. Německý ovčák. Planeta zvířat. Moravská typografie, 4:(4–5). ISSN 1211-4634.

Vacková, D. 2008. Historie malinoise. [cit. 2008-10-6], available <www.bergerbelge.cz>.

Vala, R. 2007. Nebojme se mondioringu. Psí sporty. Czech Press Group a.s.. Ústí nad Labem. 3:(23–24), ISSN 1802-1867.

Vyskočilová, D. 2008. Flyball. [cit. 2008-10-6], available <www.propage.cz/flyball>.

Wagner, R. 1996. Malinois. Svět psů. Minerva Praha. 1:(38 – 39). ISSN 1802-1867.

Wedding, B. 2004. Belgian malinois vs. German shepherd.[cit 2012-1-5], available <www.kwintessential.co.uk./articles/belgium/>.

Wegmannová, A. 2003. Jak trávit volný čas se psem. Knižní klub Praha. p.123, ISBN 80-242-0947-0.

DEW-Animal House, Faculty of Agrobiology, Food and Natural Resources, Czech University of Life Sciences Prague, Kamýcká 129, 16521, Praha 6 – Suchdol, Czech Republic

SUMMARY

This work is compilation of diploma and bachelor work and summarized the history and present of breeds Belgian Shepherd Malinois and German Shepherd. It shows formation of both breeds and their development. It introduces with conditions in individual clubs and their widespread use of working, especially in dogs sports. Belgian Shepherd Malinois is dog lively and intelligent. It's breed with bursting vitality, with which it plunges into all activities. His extension work is the same as another reference of German Shepherd dog - police use, rescue and sports training, where his vigor and vitality is interesting. However it is breed less adaptable to change handlers. The most popular breeds worldwide is German shepherd. It is also the most popular working dog and one of least demanding breeds for breeding, is very resistant and flexible. The German Shepherd has excellent qualities of character, which since the beginning of his recognition placed in front of all external qualities. It is most adaptable to change handler and environment so its use is in police and other armed forces, guide and assistance dog, herding dog, rescue and avalanche dog, sporting dog, family dog.

Statistically, two breeds were evaluated according to IPO - World Cup/Championship and the Czech IPO Championship and compared scores in disciplines - obedience, defense, track and total score from 2003 to 2011 nine years period. It was found out that Belgian Shepherd Dog received higher scores than German Shepherd in WC, although his representation as a breed was not very significant. The performance in Czech Republic of both breeds were more balanced, but the overall results except for the years 2006, 2007 and 2008 received higher scores for Belgian Shepherd Malinois. Belgian Shepherd Malinois excels in obedience and defense, German Shepherd received higher scores and excels in track.

Keywords: breeds, Belgian shepherd dog, German shepherd dog, IPO, comparison

I want morebooks!

Buy your books fast and straightforward online - at one of the world's fastest growing online book stores! Environmentally sound due to Print-on-Demand technologies.

Buy your books online at
www.get-morebooks.com

Kaufen Sie Ihre Bücher schnell und unkompliziert online – auf einer der am schnellsten wachsenden Buchhandelsplattformen weltweit! Dank Print-On-Demand umwelt- und ressourcenschonend produziert.

Bücher schneller online kaufen
www.morebooks.de

VDM Verlagsservicegesellschaft mbH
Heinrich-Böcking-Str. 6-8
D - 66121 Saarbrücken
Telefax: +49 681 93 81 567-9
info@vdm-vsg.de
www.vdm-vsg.de

www.ingramcontent.com/pod-product-compliance
Lightning Source LLC
Chambersburg PA
CBHW031541210526
45464CB00003B/1092